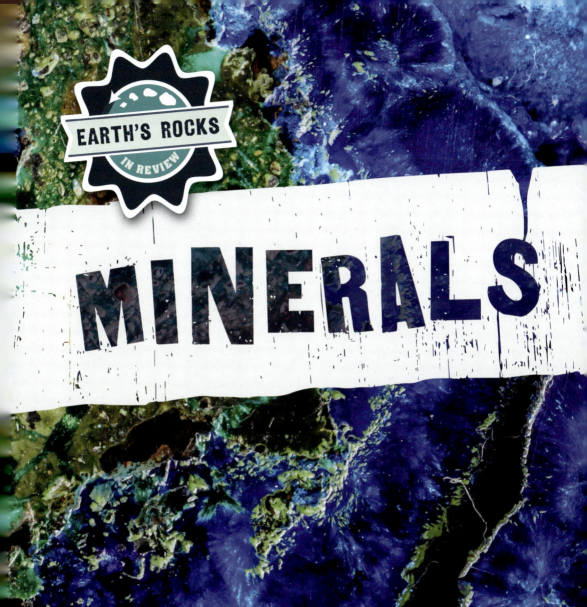

EARTH'S ROCKS IN REVIEW

MINERALS

By Anna McDougal

Please visit our website, www.enslow.com. For a free color catalog of all our high-quality books, call toll free 1-800-398-2504 or fax 1-877-980-4454.

Library of Congress Cataloging-in-Publication Data
Names: McDougal, Anna, author.
Title: Minerals / Anna McDougal.
Description: Buffalo, NY : Enslow Publishing, [2024] | Series: Earth's rocks in review | Includes bibliographical references and index.
Identifiers: LCCN 2023031098 | ISBN 9781978537958 (library binding) | ISBN 9781978537941 (paperback) | ISBN 9781978537965 (ebook)
Subjects: LCSH: Minerals–Juvenile literature. | Natural resources–Juvenile literature.
Classification: LCC QE365.2 .M3824 2024 | DDC 549–dc23/eng/20230630
LC record available at https://lccn.loc.gov/2023031098

Published in 2024 by
Enslow Publishing
2544 Clinton Street
Buffalo, NY 14224

Copyright © 2024 Enslow Publishing

Portions of this work were originally authored by Kristen Rajczak Nelson and published as *What Are Minerals?* All new material in this edition authored by Anna McDougal.

Designer: Claire Wrazin
Editor: Caitie McAneney

Photo credits: Cover, p. 1 Minakryn Ruslan/Shutterstock.com; series art (title & heading background shape) cddesign.co/Shutterstock.com; series art (dark stone background) Somchai kong/Shutterstock.com; series art (white stone header background) Madredus/Shutterstock.com; series art (light stone background) hlinjue/Shutterstock.com; series art (learn more stone background) MaraZe/Shutterstock.com; p. 5 dekazigzag/Shutterstock.com; p. 7 beboy/Shutterstock.com; p. 9 Yes058 Montree Nanta/Shutterstock.com; pp. 11, 17 (bottom) Sebastian Janicki/Shutterstock.com; pp. 11, 19 arrows Elina Li/Shutterstock.com; p. 13 BGStock72/Shutterstock.com; pp. 15, 30 (atomic structure) Kuttelvaserova Stuchelova/Shutterstock.com; p. 17 (top) Stefan Malloch/Shutterstock.com; pp. 19, 30 (color) Albert Russ/Shutterstock.com; pp. 21 (top), 25 (bottom), 30 (luster) RHJPhtotos/Shutterstock.com; pp. 21 (bottom), 30 (cleavage) Levon Avagyan/Shutterstock.com; pp. 23 (top and bottom), 30 (streak, hardness) Michael LaMonica/Shutterstock.com; p. 25 (top) Oleksii Holikov/Shutterstock.com; p. 27 (top) Byjeng/Shutterstock.com, p. 27 (bottom) Pixel-Shot/Shutterstock.com; p. 29 Kzenon/Shutterstock.com.

All rights reserved. No part of this book may be reproduced in any form without permission in writing from the publisher, except by a reviewer.

Printed in the United States of America

Some of the images in this book illustrate individuals who are models. The depictions do not imply actual situations or events.

CPSIA compliance information: Batch #CWENS24: For further information, contact Enslow Publishing at 1-800-398-2504.

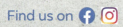

CONTENTS

What Are Minerals? 4
The Making of Minerals 6
Awesome Atoms 10
Crystals Up Close 14
Luster and Color 18
Break It Down 20
Streak and Hardness 22
Finding Minerals 24
Gemstones 26
Lab-Made Minerals 28
Properties of Minerals 30
Glossary 31
For More Information 32
Index 32

Words in the glossary appear in **bold** the first time they are used in the text.

WHAT ARE MINERALS?

Imagine the earth beneath you. It is made up of minerals. Minerals are **inorganic** matter. A mineral is a solid, nonliving thing found in nature. It has a certain composition, or makeup. This composition must be the same throughout.

LEARN MORE

Rocks, sands, and soils are all made up of minerals.

THE MAKING OF MINERALS

Minerals have been around for billions of years. All rocks are made up of one or more minerals. Most minerals form when atoms float freely in a fluid and then come together when the liquid cools.

LEARN MORE

More than 4,000 minerals occur naturally on Earth.

7

Minerals can form from other minerals. This happens when conditions change, such as rising heat. Other times, minerals **dissolve** in water and are carried somewhere. Once they reach their resting spot, they form new minerals.

LEARN MORE

Minerals grow together or are hardened together to become rocks. You can see the **grains** of different minerals in some rocks.

AWESOME ATOMS

Each mineral has its own special pattern of atoms. The pattern repeats over and over, giving the mineral a certain atomic **structure**. Atomic structure is how minerals are grouped. There are eight main groups of minerals, including silicates, sulfides, and halides.

quartz

LEARN MORE

The mineral quartz is made up of one silicon atom and two oxygen atoms.

11

Scientists use special tools to **magnify** minerals. They look at a mineral's elements and atomic structure. The pattern of atoms and elements give minerals certain **properties** that allow scientists to **identify** them. Properties include the mineral's crystal shape.

LEARN MORE

A crystal is a small piece of something that has many sides and is created when a substance becomes a solid.

CRYSTALS UP CLOSE

The look of a mineral's crystals can be used to identify minerals. Crystals have different shapes and patterns. For example, salt crystals are shaped like cubes, which helps scientists identify salt. There are seven main groups of crystal types.

salt crystals

LEARN MORE

Crystals only "grow" in the right conditions. They need enough heat and sometimes thousands of years to form.

15

Have you ever seen quartz? Usually, it is colorless. But sometimes it is pink, purple, or brown. That's because a **trace** of other elements can make crystals of a mineral look different. This is called an impurity.

LEARN MORE

Rose quartz looks pink because it has trace amounts of the elements iron, manganese, or titanium.

rose quartz

brown quartz

LUSTER AND COLOR

Looking at atomic structure is sometimes only possible with a powerful **microscope**. Luckily, some mineral properties are easier to see. One example is luster. That's how a mineral reflects, or gives back, light. Some minerals are shiny and some are dull.

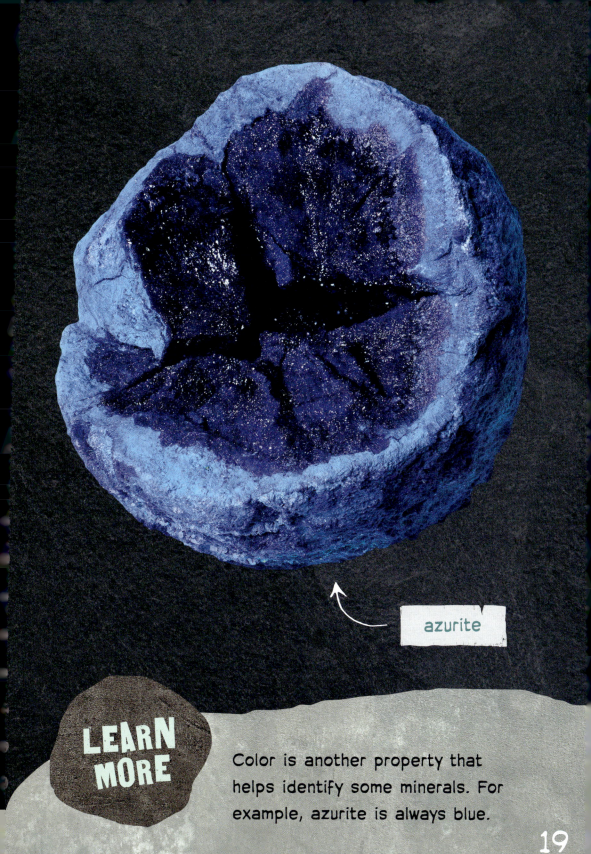

azurite

LEARN MORE

Color is another property that helps identify some minerals. For example, azurite is always blue.

BREAK IT DOWN

Another mineral property is cleavage, or how a mineral breaks. Some minerals break along certain lines. If a crystal breaks along a smooth, flat plane, it's said to have cleavage. Cleavage describes how easily and how smoothly a mineral breaks.

LEARN MORE

Diamonds break easily and smoothly. They have perfect cleavage in four different places. That's why people like to make them into **jewelry!**

natural diamond

diamond ring

STREAK AND HARDNESS

Scientists also look at a mineral's streak and hardness to identify it. Streak is the color of a mineral when it's smashed to powder. Scientists test hardness by seeing how easily a mineral is scratched by other minerals.

LEARN MORE

To test streak, a mineral is rubbed on a piece of tile. The color of the powder it leaves helps identify the mineral.

mineral streak test

mineral hardness test

23

FINDING MINERALS

Minerals are found everywhere! People dig rare, or uncommon, minerals out of Earth. Some are found deep underground, and big machines must be used to dig them up. This is called mining. Companies mine for useful minerals, such as copper and iron ore.

LEARN MORE

Minerals are used to make many things. Copper is used to make pipes and wires.

copper mine

copper

GEMSTONES

Jewelry often includes gemstones, or pieces of minerals cut to look a certain way. They usually have a beautiful color and reflect light well. Gemstones include diamonds, emeralds, sapphires, and rubies. People pay a lot of money for these minerals!

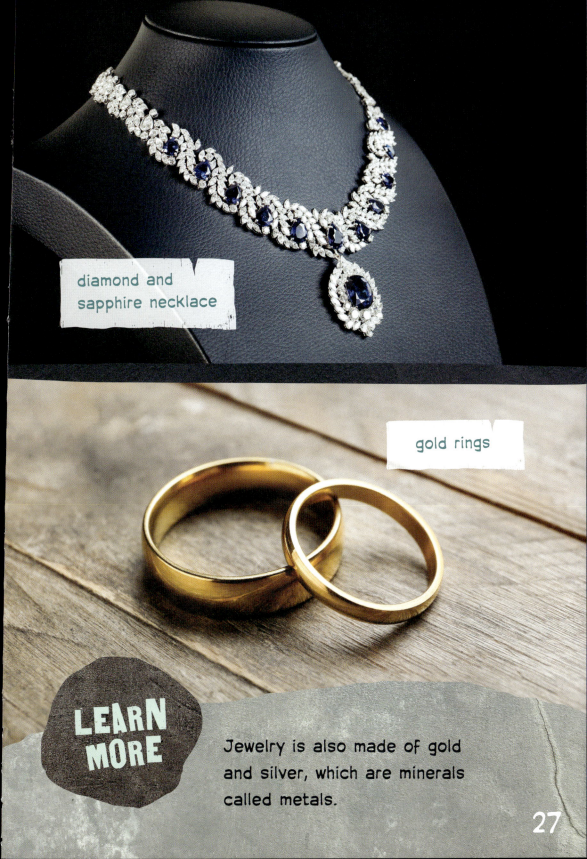

diamond and sapphire necklace

gold rings

LEARN MORE

Jewelry is also made of gold and silver, which are minerals called metals.

27

LAB-MADE MINERALS

Some human-made minerals are created in a lab. They're called synthetic minerals. They're not true minerals because they didn't occur in nature. Synthetic gems have been created since the late 1800s, and they look like true gems.

LEARN MORE

Synthetic gems are better for the environment, or natural world, because they don't have to be mined.

29

LUSTER

COLOR

HARDNESS

PROPERTIES OF MINERALS

CLEAVAGE

ATOMIC STRUCTURE

STREAK

dissolve: To mix completely into a liquid.

element: Matter that is pure and has no other type of matter in it.

grain: An individual particle that can be found in rocks. Grain size is often an important property for identifying rocks.

identify: To find out the name or features of something.

inorganic: Being made of matter that doesn't come from plants or animals.

jewelry: Pieces of metal, often holding gems, worn on the body.

magnify: To make something appear larger.

microscope: A tool used to view very small objects so they can be seen much larger and more clearly.

property: Something used to tell a member of a group or one thing from another.

structure: The way something is arranged, or set up.

trace: A very small amount of something.

FOR MORE INFORMATION

BOOKS
Farndon, John. *Rocks and Minerals*. New York, NY: Scholastic Press, 2022.

Pettiford, Rebecca. *Minerals*. Minneapolis, MN: Jump!, Inc, 2024.

WEBSITE
Gemstones
kids.nationalgeographic.com/science/topic/gemstones
Explore more fun facts about gemstones.

Publisher's note to educators and parents: Our editors have carefully reviewed this website to ensure it is suitable for students. Many websites change frequently, however, and we cannot guarantee that a site's future contents will continue to meet our high standards of quality and educational value. Be advised that students should be closely supervised whenever they access the internet.

INDEX

atomic structure, 10, 12, 18, 30

cleavage, 20, 30

color, 16, 19, 30

crystal, 12, 13, 14, 15, 16, 20, 30

elements, 11, 12, 16

gemstones, 26, 28, 29

hardness, 22, 23

jewelry, 20, 26

luster, 18, 30

metals, 27

mining, 24

streak, 22, 23, 30

trace elements, 16